Drug Development

Many drugs are based on natural products. Aspirin (above) is a chemical derivative of a compound isolated from willow bark (near left). Extracts of willow bark had been long known to have medicinal properties. The active compound was isolated, modified, and, beginning in 1899, packaged for consumers (far left). [*Far left:* Used with permission of Bayer Corporation. *Near left:* Image Ideas/Picture Quest.]

The development of drugs represents one of the most important interfaces between biochemistry and medicine. In most cases, drugs act by binding to specific receptors or enzymes and inhibiting, or otherwise modulating, their activities. Thus, knowledge of these molecules and the pathways in which they participate is crucial to drug development. An effective drug is much more than a potent modulator of its target, however. Drugs must be readily administered to patients, preferably as small tablets taken orally, and must survive within the body long enough to reach their targets. Furthermore, to prevent unwanted physiological effects, drugs must not modulate the properties of biomolecules other than the target molecules. These requirements tremendously limit the number of compounds that have the potential to be clinically useful drugs.

Drugs have been discovered by two, fundamentally opposite, approaches (Figure 1). The first approach identifies a substance that has a desirable physiological consequence when administered to a human being, to an appropriate animal, or to cells. Such substances can be discovered by serendipity, by the fractionation of plants or other materials known to have medicinal properties, or by screening natural products or other "libraries" of compounds. In this approach, a biological effect is

Outline

1 The Development of Drugs Presents Huge Challenges

2 Drug Candidates Can Be Discovered by Serendipity, Screening, or Design

3 The Analysis of Genomes Holds Great Promise for Drug Discovery

4 The Development of Drugs Proceeds Through Several Stages

Drug Development

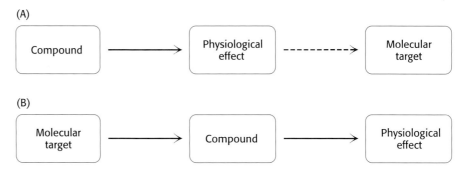

Figure 1 **Two paths to drug discovery.** (A) A compound is discovered to have a desirable physiological effect. The molecular target can be identified in a separate step as needed. (B) A molecular target is selected first. Drug candidates that bind to the target are identified and then examined for their physiological effects.

known before the molecular target is identified. The mode of action of the substance is only later identified after substantial additional work. The second approach begins with a known molecular target. Compounds are sought, either by screening or by designing molecules with desired properties, that bind to the target molecule and modulate its properties. Once such compounds are available, scientists can explore their effects on appropriate cells or organisms. Many unexpected results may be encountered in this process as the complexity of biological systems reveals itself.

In this chapter, we explore the science of pharmacology. We examine a number of case histories that illustrate drug development—including many of its concepts, methods, and challenges. We then see how the concepts and tools from genomics are influencing approaches to drug development. We conclude the chapter with a summary of the stages along the way to developing a drug.

> **Pharmacology**
> The science that deals with the discovery, chemistry, composition, identification, biological and physiological effects, uses, and manufacture of drugs.

1 The Development of Drugs Presents Huge Challenges

Many compounds have significant effects when taken into the body, but only a very small fraction of them have the potential to be useful drugs. A foreign compound, not adapted to its role in the cell through long evolution, must have a range of special properties to function effectively without causing serious harm. We next review some of the challenges faced by drug developers.

Drug Candidates Must Be Potent Modulators of Their Targets

Most drugs bind to specific proteins, usually receptors or enzymes, within the body. To be effective, a drug needs to bind a sufficient number of its target proteins when taken at a reasonable dose. One factor in determining drug effectiveness is the strength of binding, often governed by the principles of binding, related to the Michaelis-Menten model introduced in Chapter 8.

A molecule that binds to some target molecule is often referred to as a *ligand*. A ligand-binding curve is shown in Figure 2. Ligand molecules occupy progressively more target binding sites as ligand concentration increases until essentially all of the available sites are occupied. The tendency

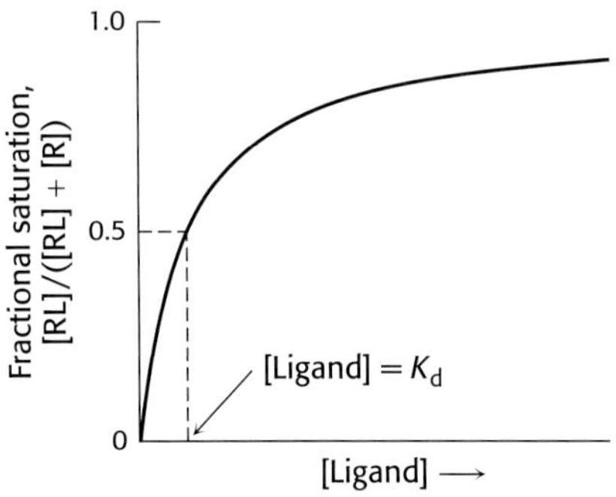

Figure 2 **Ligand binding.** The titration of a receptor, R, with a ligand, L, results in the formation of the complex RL. In uncomplicated cases, the binding reaction follows a simple saturation curve. Half of the receptors are bound to ligand when the ligand concentration equals the dissociation constant, K_d, for the RL complex.

of a ligand to bind to its target is measured by the *dissociation constant*, K_d, defined by the expression

$$K_d = [R][L]/[RL]$$

where [R] is the concentration of the receptor, [L] is the concentration of the ligand, and [RL] is the concentration of the receptor–ligand complex. The dissociation constant is a measure of the strength of the interaction between the drug candidate and the target; the lower the value, the stronger the interaction. The concentration of free ligand at which one-half of the binding sites are occupied equals the dissociation constant, as long as the concentration of binding sites is substantially less than the dissociation constant.

Many complicating factors are present under physiological conditions. Many drug targets also bind ligands normally present in tissues; these ligands and the drug candidate compete for binding sites on the target. We encountered this situation when we considered competitive inhibitors in Chapter 8. Suppose that the drug target is an enzyme and the drug candidate is a competitive inhibitor. The concentration of the drug candidate necessary to inhibit the enzyme effectively will depend on the physiological concentration of the enzyme's normal substrate (Figure 3). The higher the concentration of the endogenous substrate, the higher the concentration of drug candidate needed to inhibit the enzyme to a given extent. This effect of substrate concentration is expressed by the *apparent dissociation constant*, K_d^{app}. The apparent dissociation constant is given by the expression

$$K_d^{app} = K_d(1 + [S]/K_M)$$

where [S] is the concentration of substrate and K_M is the Michaelis constant for the substrate. Note that, for an enzyme inhibitor, the dissociation constant, K_d, is often referred to as the *inhibition constant*, K_i.

In many cases, more complicated biological assays (rather than direct enzyme or binding assays) are used to examine the potency of drug candidates. For example, the fraction of bacteria killed might indicate the potency of a potential antibiotic. In these cases, values such as EC_{50} are used. EC_{50} is the concentration of drug candidate required to elicit 50% of the maximal biological response (Figure 4). Similarly, EC_{90} is the concentration required to achieve 90% of the maximal response. In the example of an antibiotic, EC_{90} would be the concentration required to kill 90% of bacteria exposed to the drug. For inhibitors, the corresponding terms IC_{50} and IC_{90} are often used to describe the concentrations of the inhibitor required to reduce a response to 50% or 90% of its value in the absence of inhibitor, respectively.

These values are measures of the potency of a drug candidate in modulating the activity of the desired biological target. To prevent unwanted effects, often called *side effects*, ideal drug candidates should not bind biomolecules other than the target to any appreciable extent. Developing such a drug can be quite challenging, particularly if the drug target is a member of a large family of evolutionarily related proteins. The degree of specificity can be described in terms of the ratio of the K_d values for the binding of the drug candidate to any other molecules to the K_d value for the binding of the drug candidate to the desired target.

Drugs Must Have Suitable Properties to Reach Their Targets

Thus far, we have focused on the ability of molecules to act on specific target molecules. However, an effective drug must also have other characteristics.

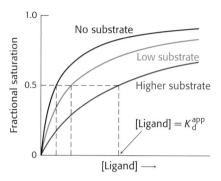

Figure 3 Inhibitors compete with substrates for binding sites. These binding curves give results for an inhibitor binding to a receptor in the absence of substrate and in the presence of increasing concentrations of substrate.

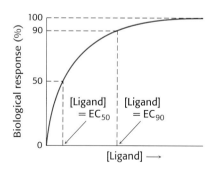

Figure 4 Effective concentrations. The concentration of a ligand required to elicit a biological response can be quantified in terms of EC_{50}, the concentration required to give 50% of the maximum response, and EC_{90}, the concentration required to give 90% of the maximum response.

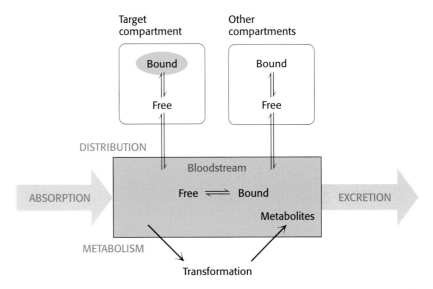

Figure 5 Absorption, distribution, metabolism, and excretion. The concentration of a compound at its target site (yellow) is affected by the extents and rates of absorption, distribution, metabolism, and excretion.

It must be easily administered and must reach its target at sufficient concentration to be effective. A drug molecule encounters a variety of obstacles on its way to its target, related to its absorption, distribution, metabolism, and excretion after it has entered the body. These processes are interrelated to one another as summarized in Figure 5. Taken together, a drug's ease of absorption, distribution, metabolism, and excretion are often referred to as *ADME* (pronounced "add-me") properties.

Administration and absorption. Ideally, a drug can be taken orally as a small tablet. An orally administered active compound must be able to survive the acidic conditions in the gut and then be absorbed through the intestinal epithelium. Thus, the compound must be able to pass through cell membranes at an appreciable rate. Larger molecules such as proteins cannot be administered orally, because they often cannot survive the acidic conditions in the stomach and, if they do, are not readily absorbed. Even many small molecules are not absorbed well, because, for example, if they are too polar they do not pass through cell membranes readily. The ability to be absorbed is often quantified in terms of the *oral bioavailability*. This quantity is defined as the ratio of the peak concentration of a compound given orally to the peak concentration of the same dose injected directly into the bloodstream. Bioavailability can vary considerably from species to species so results from animal studies may be difficult to translate to human beings. Despite this variability, some useful generalizations have been made. One powerful set is *Lipinski's rules*.

Lipinski's rules tell us that poor absorption is likely when

1. the molecular weight is greater than 500.
2. the number of hydrogen-bond donors is greater than 5.
3. the number of hydrogen-bond acceptors is greater than 10.
4. the partition coefficient [measured as $\log(P)$] is greater than 5.

The partition coefficient is a way to measure the tendency of a molecule to dissolve in membranes, which correlates with its ability to dissolve in organic solvents. It is determined by allowing a compound to equilibrate between water

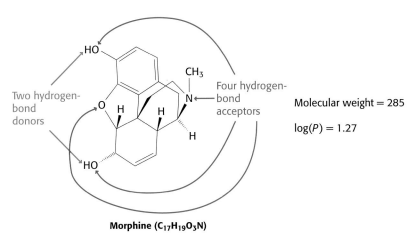

Figure 6 Lipinski's rules applied to morphine. Morphine satisfies all of Lipinski's rules and has an oral bioavailability in human beings of 33%.

and an organic phase, n-octanol. The $\log(P)$ value is defined as $\log_{10}$ of the ratio of the concentration of a compound in n-octanol to the concentration of the compound in water. For example, if the concentration of the compound in the n-octanol phase is 100 times that in the aqueous phase, then $\log(P)$ is 2.

Morphine, for example, satisfies all of Lipinski's rules and has moderate bioavailability (Figure 6). A drug that violates one or more of these rules may still have satisfactory bioavailability. Nonetheless, these rules serve as guiding principles for evaluating new drug candidates.

Distribution. Compounds taken up by intestinal epithelial cells can pass into the bloodstream. However, hydrophobic compounds and many others do not freely dissolve in the bloodstream. These compounds bind to proteins, such as albumin (Figure 7), that are abundant in the serum and by this means are carried everywhere that the bloodstream goes.

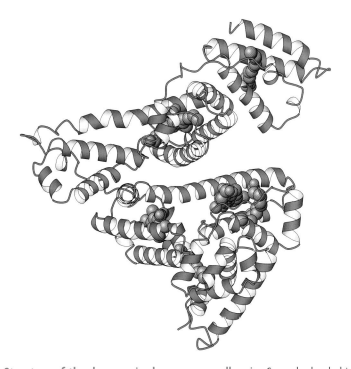

Figure 7 Structure of the drug carrier human serum albumin. Seven hydrophobic molecules (in red) are shown bound to the molecule.

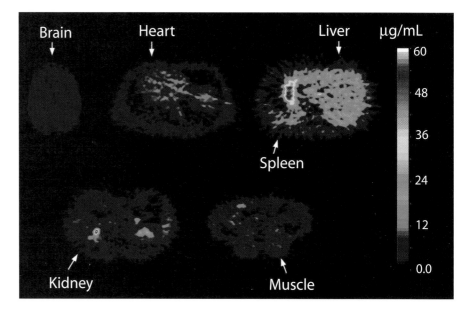

Figure 8 Distribution of the drug fluconazole. Once taken in, compounds distribute themselves to various organs within the body. The distribution of the antifungal agent fluconazole has been monitored through the use of positron emission tomography (PET) scanning. These images were taken of a healthy human volunteer 90 minutes after injection of a dose of 5 mg kg^{-1} of fluconazole containing trace amounts of fluconazole labeled with the positron-emitting isotope ^{18}F. [From A. J. Fischman et al., *Antimicrob. Agents Chemother.* 37(1993): 1270–1277.]

When a compound has reached the bloodstream, it is distributed to different fluids and tissues, which are often referred to as *compartments*. Some compounds are highly concentrated in their target compartments, either by binding to the target molecules themselves or by other mechanisms. Other compounds are distributed more widely (Figure 8). An effective drug will reach the target compartment in sufficient quantity; the concentration of the compound in the target compartment is reduced whenever the compound is distributed into other compartments.

Some target compartments are particularly hard to reach. Many compounds are excluded from the central nervous system by the *blood–brain barrier,* the tight junctions between endothelial cells that line blood vessels within the brain and spinal cord.

Metabolism and excretion. A final challenge to a potential drug molecule is to evade the body's defenses against foreign compounds. Such compounds (often called *xenobiotic compounds*) are often released from the body in the urine or stool, often after having been metabolized somehow— degraded or modified—to aid in excretion. This *drug metabolism* poses a considerable threat to drug effectiveness because the concentration of the desired compound decreases as it is metabolized. Thus, a rapidly metabolized compound must be administered more frequently or at higher doses.

Two of the most common pathways in xenobiotic metabolism are *oxidation* and *conjugation*. Oxidation reactions can aid excretion in at least two ways: by increasing water solubility, and thus ease of transport, and by introducing functional groups that participate in subsequent metabolic steps. These reactions are often promoted by cytochrome P450 enzymes in the liver (Section 26.4.3). The human genome encodes more than 50 different P450 isozymes, many of which participate in xenobiotic metabolism. A typical reaction catalyzed by a P450 isozyme is the hydroxylation of ibuprofen (Figure 9).

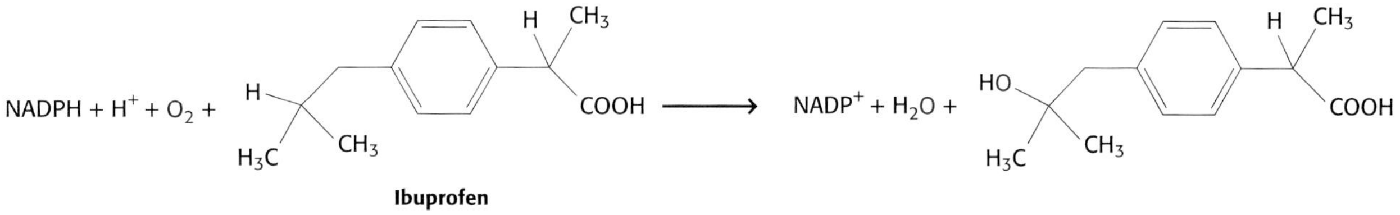

Conjugation is the addition of particular groups to the xenobiotic compound. Common groups added are glutathione (Section 20.5), glucuronic acid, and sulfate (Figure 10). The addition often increases water solubility and provides labels that can be recognized to target excretion. Examples of conjugation include the addition of glutathione to the anticancer drug cyclophosphamide, the addition of glucuronidate to the analgesic morphine, and the addition of a sulfate group to the hair-growth stimulator minoxidil.

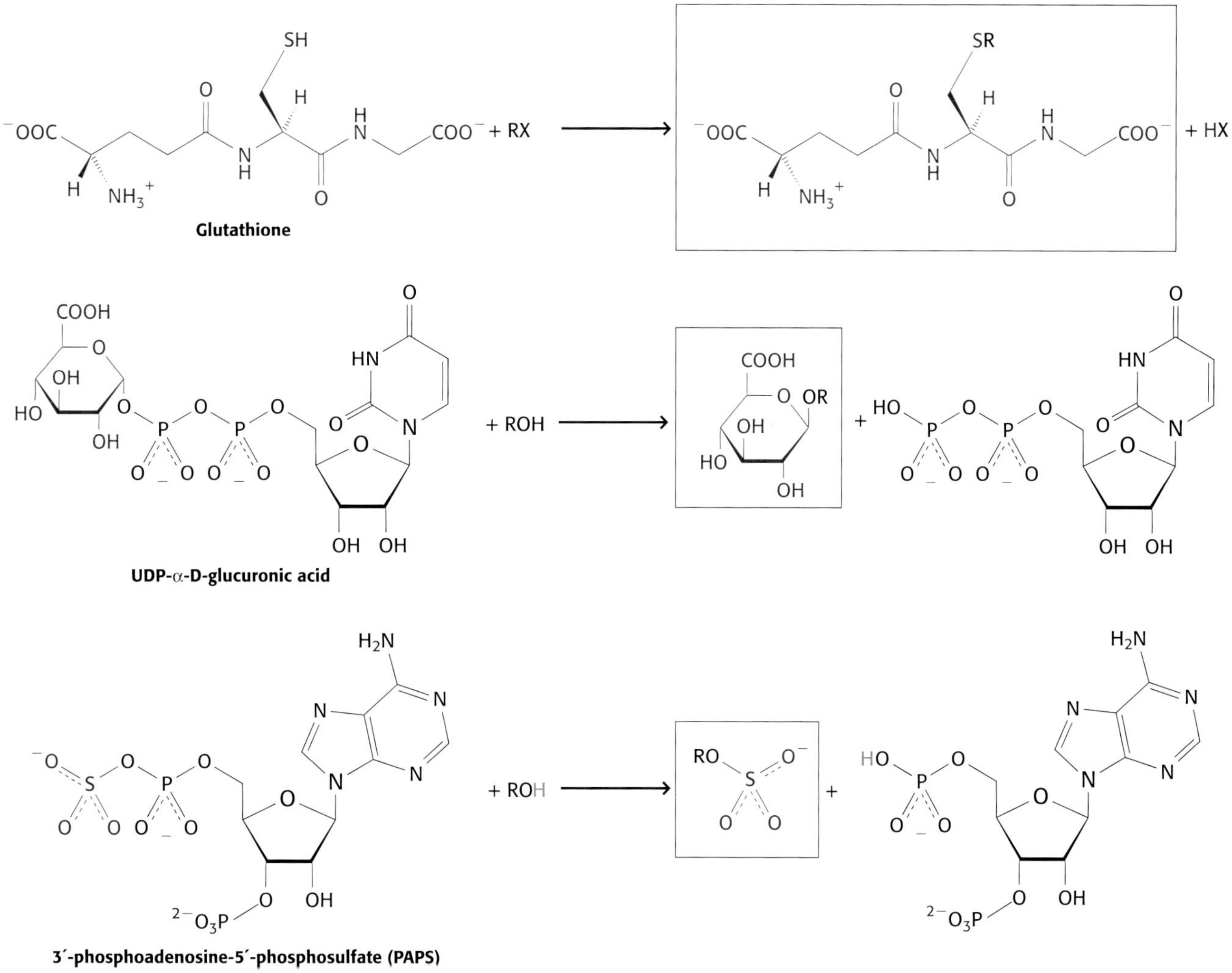

Figure 9 P450 conversion of ibuprofen. Cytochrome P450 isozymes, primarily in the liver, catalyze xenobiotic metabolic reactions such as hydroxylation. The reaction introduces an oxygen atom derived from molecular oxygen.

Figure 10 Conjugation reactions. Compounds that have appropriate groups are often modified by conjugation reactions. Such reactions include the addition of glutathione (top), glucuronic acid (middle), or sulfate (bottom). The conjugated product is shown boxed.

Cyclophosphamide-glutathione conjugate **Morphine glucuronidate** **Minoxidil sulfate**

Interestingly, the sulfation of minoxidil produces a compound that is more active in stimulating hair growth than is the unmodified compound. Thus, the metabolic products of a drug, though usually less active than the drug, can sometimes be more active.

Note that an oxidation reaction often precedes conjugation because the oxidation reaction can generate hydroxyl and other groups to which groups such as glucuronic acid can be added. The oxidation reactions of xenobiotic compounds are often referred to as *phase I transformations,* and the conjugation reactions are referred to as *phase II transformations.* These reactions take place primarily in the liver. Because blood flows from the intestine directly to the liver through the portal vein, xenobiotic metabolism often alters drug compounds before they ever reach full circulation. This *first-pass metabolism* can substantially limit the availability of compounds taken orally.

After compounds have entered the bloodstream, they can be removed from circulation and excreted from the body by two primary pathways. First, they can be absorbed through the kidneys and excreted in the urine. In this process, the blood passes through *glomeruli,* networks of fine capillaries in the kidney that act as filters. Compounds with molecular weights less than approximately 60,000 pass though the glomeruli into the kidney. Many of the water molecules, glucose molecules, nucleotides, and other low-molecular-weight compounds that pass through the glomeruli are reabsorbed into the bloodstream, either by transporters that have broad specificities or by the passive transfer of hydrophobic molecules through membranes. Drugs and metabolites that pass through the first filtration step and are not reabsorbed are excreted.

Second, compounds can be actively transported into bile, a process that takes place in the liver. After concentration, bile flows into the intestine. In the intestine, the drugs and metabolites can be excreted through the stool, reabsorbed into the bloodstream, or further degraded by digestive enzymes. Sometimes, compounds are recycled from the bloodstream into the intestine and back into the bloodstream, a process referred to as *enterohepatic cycling* (Figure 11). This process can significantly decrease the rate of excretion of some compounds because they escape from an excretory pathway and reenter the circulation.

The kinetics of compound excretion is often complex. In some cases, a fixed percentage of the remaining compound is excreted over a given period of time (Figure 12). This pattern of excretion results in exponential loss of the compound from the bloodstream that can be characterized by a half-life ($t_{1/2}$). The half-life is the fixed period of time required to eliminate 50% of the remaining compound. It is a measure of how long an effective concentration of the compound remains in the system after administration. As such, the half-life is a major factor in determining how often a drug must be taken. A drug with a long half-life might need to be taken only once per day, whereas a drug with a short half-life might need to be taken three or four times per day.

Toxicity Can Limit Drug Effectiveness

An effective drug must not be so toxic that it seriously harms the person who takes it. A drug may be toxic for any of several reasons. First, it may modulate the target molecule itself *too* effectively. For example, the presence of too much of the anticoagulant drug coumadin can result in dangerous, uncontrolled bleeding and death. Second, the compound may modulate the properties of proteins that are distinct from, but related to, the target molecule itself. Compounds that are directed to one member of a family of enzymes or receptors often bind to other family members. For example, an antiviral drug directed against viral proteases may be toxic if it also inhibits proteases normally present in the body such as those that regulate blood pressure.

A compound may also be toxic if it modulates the activity of a protein unrelated to its intended target. For example, many compounds block ion channels such as the potassium channel HERG (the human homolog of a *Drosophila* channel found in a mutant termed "ether-a-go-go"), causing disturbances of the heartbeat. To avoid cardiac side effects, many compounds are screened for their ability to block such channels.

Finally, even if a compound is not itself toxic, its metabolic by-products may be. Phase I metabolic processes can generate damaging reactive groups in products. An important example is liver toxicity observed with large doses of acetaminophen (Figure 13). A particular cytochrome P450 isozyme oxidizes acetaminophen to *N*-acetyl-*p*-benzoquinone imine. The resulting compound is conjugated to glutathione. With large doses, however, the liver concentration of glutathione drops dramatically, and the liver is no longer able to protect itself from this reactive compound and others. Initial symptoms of excessive acetaminophen include nausea and vomiting. Within 24 to 48 hours, symptoms of liver failure may appear. Acetaminophen poisoning accounts for about 35% of cases of severe liver failure in the United States. A liver transplant is often the only effective treatment.

The toxicity of a drug candidate can be described in terms of the *therapeutic index*. This measure of toxicity is determined through animal tests, usually with mice or rats. The therapeutic index is defined as the ratio of the dose of a compound that is required to kill one-half of the animals (referred to as the LD_{50} for "lethal dose") to a comparable measure of the effective dose, usually the EC_{50}. Thus, if the therapeutic index is 1000,

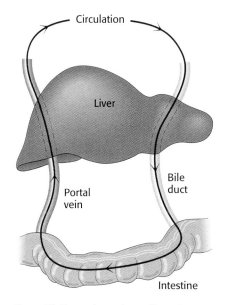

Figure 11 **Enterohepatic cycling.** Some drugs can move from the blood circulation to the liver, into the bile, into the intestine, to the liver, and back into circulation. This cycling decreases the rate of drug excretion.

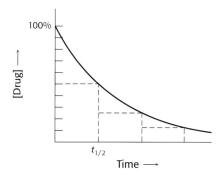

Figure 12 **Half-life of drug excretion.** In the case shown, the concentration of a drug in the bloodstream decreases to one-half of its value in a period of time, $t_{1/2}$, referred to as its half-life.

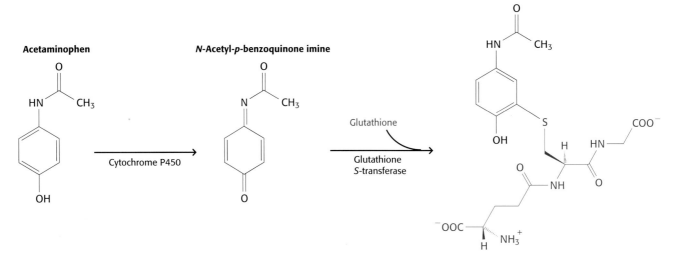

Figure 13 Acetaminophen toxicity. A minor metabolic product of acetaminophen is N-acetyl-p-benzoquinone imine. This metabolite is conjugated to glutathione. Large doses of acetaminophen can deplete liver glutathione stores.

then lethality is significant only when 1000 times the effective dose is administered. Analogous indices can provide measures of toxicity less severe than lethality.

Many compounds have favorable properties in vitro, yet fail when administered to a living organism because of difficulties with ADME and toxicity. Expensive and time-consuming animal studies are required to verify that a drug candidate is not toxic, yet differences between animal species in their response can confound decisions about moving forward with a compound toward human studies. One hope is that, with more understanding of the biochemistry of these processes, scientists can develop computer-based models to replace or augment animal tests. Such models would need to accurately predict the fate of a compound inside a living organism from its molecular structure or other properties that are easily measured in the laboratory without the use of animals.

2 Drug Candidates Can Be Discovered By Serendipity, Screening, or Design

Traditionally, many drugs were discovered by serendipity, or chance observation. More recently, drugs have been discovered by screening collections of natural products or other compounds for compounds that have desired medicinal properties. Alternatively, scientists have designed specific drug candidates by using their knowledge about a preselected molecular target. We will examine several examples of each of these pathways to reveal common principles.

Serendipitous Observations Can Drive Drug Development

Perhaps the most well known observation in the history of drug development is Alexander Fleming's chance observation in 1928 that colonies of the bacterium *Staphylococcus aureus* died when they were adjacent to colonies of the mold *Penicillium notatum*. Spores of the mold had landed accidentally on plates growing the bacteria. Fleming soon realized that the mold produced a substance that could kill disease-causing bacteria. This discovery led to a fundamentally new approach to the treatment of bacterial infections. Howard Flory and Ernest Chain developed a powdered form of the substance, termed penicillin, that became a widely used antibiotic in the 1940s.

The structure of this antibiotic was elucidated in 1945. The most notable feature of this structure is the four-membered β-lactam ring. This unusual feature is key to the antibacterial function of penicillin, as noted earlier (Section 8.5.5).

Three steps were crucial to fully capitalize on Fleming's discovery. First, an industrial process was developed for the production of penicillin from *Penicillium* mold on a large scale. Second, penicillin and penicillin derivatives were chemically synthesized. The availability of synthetic penicillin derivatives opened the way for scientists to explore the relations between structure and function. Many such penicillin derivatives have found widespread use in medicine. Finally, Jack Strominger and James Park independently elucidated the mode of action of penicillin in 1965 (Figure 14), as introduced in Chapter 8.

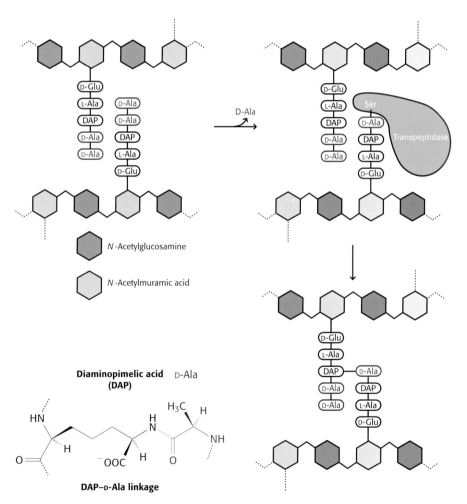

Figure 14 Mechanism of cell-wall biosynthesis. A transpeptidase enzyme catalyzes the formation of cross-links between peptidoglycan groups. Penicillin inhibits the action of the transpeptidase; so bacteria exposed to the drug have weak cell walls that are susceptible to lysis. In the case shown, the transpeptidase catalyzes the linkage of D-alanine at the end of one peptide chain to the amino acid diaminopimelic acid (DAP) on another peptide chain. The diaminopimelic acid linkage (bottom left) is found in Gram-negative bacteria such as *E. coli*. Linkages of glycine-rich peptides are found in Gram-positive bacteria.

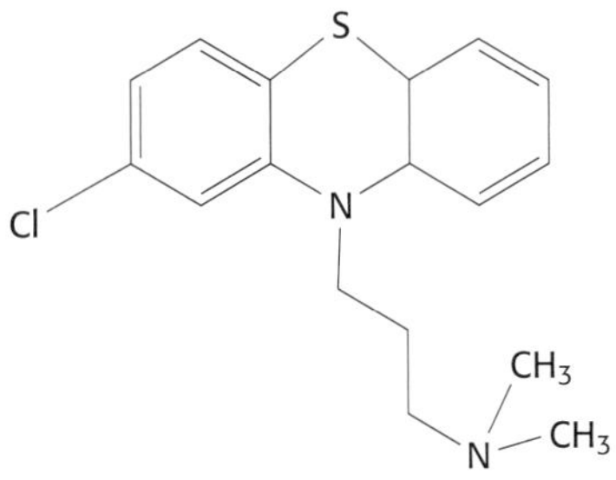

Many other drugs have been discovered by serendipitous observations. The antineuroleptic drug chlorpromazine (Thorazine®) was discovered in the course of investigations directed toward the treatment of shock in surgical patients. In 1952, French surgeon Henri Laborit noticed that, after taking the compound, his patients were remarkably calm. This observation suggested that chlorpromazine could benefit psychiatric patients, and, indeed, the drug has been used for many years to treat patients with schizophrenia and other disorders. The drug does have significant side effects, and its use has been largely superceded by more recently developed drugs.

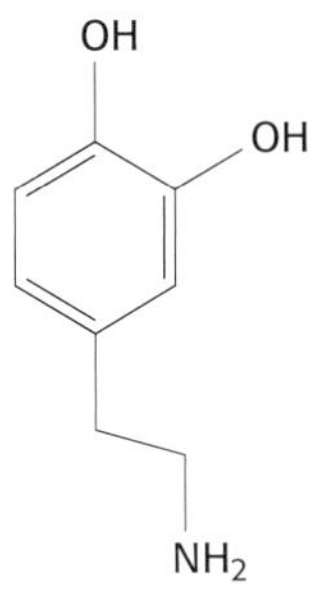

Chlorpromazine acts by binding to receptors for the neurotransmitter dopamine and blocking them (Figure 15). Dopamine D2 receptors are the targets of many other psychoactive drugs. In the search for drugs with more limited side effects, studies are undertaken to correlate drug effects with biochemical parameters such as dissociation constants and binding and release rate constants.

A more recent example of a drug discovered by chance observation is sildenafil (Viagra®). This compound was developed as an inhibitor of phosphodiesterase 5, an enzyme that catalyzes the hydrolysis of cGMP to GMP (Figure 16). The compound was intended as a treatment for hypertension and angina because cGMP plays a central role in the relaxation of smooth muscle cells in blood vessels (Figure 17). Inhibiting phosphodiesterase 5 was expected to increase the concentration of cGMP by blocking the pathway for its degradation. In the course of early clinical trials in Wales, some men reported unusual penile erections. Whether this chance observation by a few men was due to the compound or to other effects was unclear. However, the observation made some biochemical sense because smooth muscle relaxation due to increased cGMP levels had been discovered to play a role in penile erection. Subsequent clinical trials directed toward the evaluation of sildenafil for erectile dysfunction were successful. This account testifies to the importance of collecting comprehensive information from clinical-trial participants. In this case, incidental observations led to a new treatment for erectile dysfunction and a multibillion-dollar-per-year drug market.

Screening Libraries of Compounds Can Yield Drugs or Drug Leads

No drug is as widely used as aspirin. Observers at least as far back as Hippocrates (~400 B.C.) have noted the use of extracts from the bark and leaves

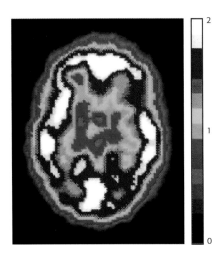

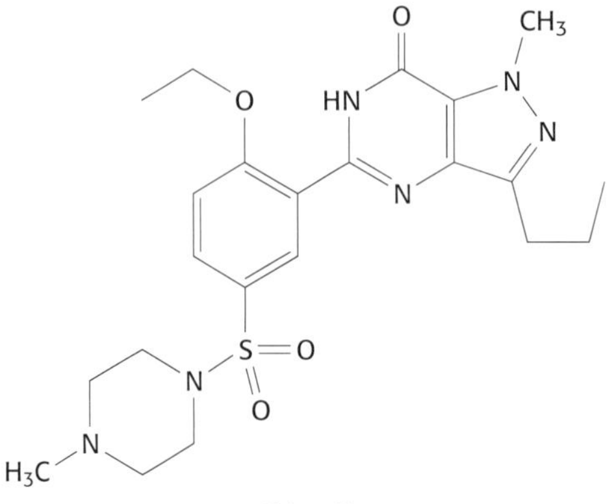

Figure 15 Chlorpromazine targets. This positron emission tomographic image shows the distribution of dopamine D2 receptors in the brain. These sites are blocked by treatment with chlorpromazine. [From C. Trichard et al., *Am. J. Psychiatry* 155 (1998): 505–508; reprinted with permission conveyed through Copyright Clearance Center, Inc.]

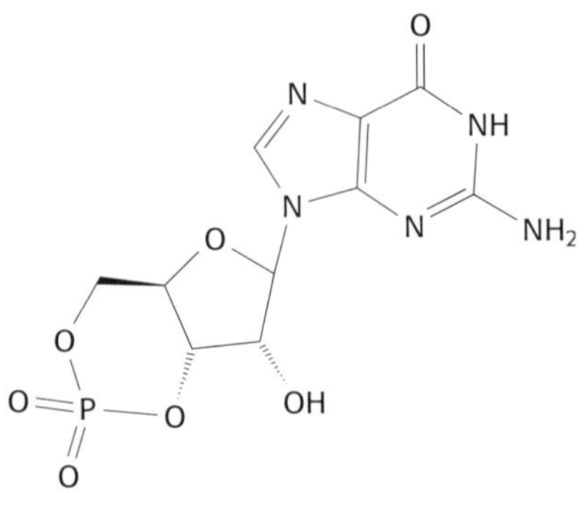

Figure 16 Sildenafil, a mimic of cGMP. Sildenafil was designed to resemble cGMP, the substrate of phosphodiesterase 5.

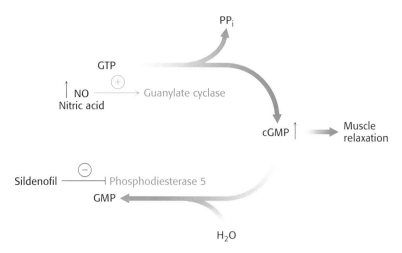

Figure 17 Muscle-relaxation pathway. Increases in NO levels stimulate guanylate cyclase, which produces cGMP. The increased cGMP concentration promotes muscle relaxation. Phosphosdiesterase 5 hydrolyzes cGMP, which lowers the cGMP concentration. The inhibition of phosphodiesterase 5 by sildenafil maintains elevated levels of cGMP.

of the willow tree for pain relief. In 1829, a mixture called *salicin* was isolated from willow bark. Subsequent analysis identified salicylic acid as the active component of this mixture. Salicylic acid was formerly used to treat pain, but this compound often irritated the stomach. Several investigators attempted to find a means to neutralize salicylic acid. Felix Hoffmann, a chemist working at the German company Bayer, developed a less-irritating derivative by treating salicylic acid with a base and acetyl chloride. This derivative, acetylsalicylic acid, was named *aspirin* from "a" for acetyl chloride, "spir" for *Spiraea ulmaria* (meadowsweet, a flowering plant that also contains salicylic acid), and "in" (a common ending for drugs). Each year, approximately 35,000 tons of aspirin are taken worldwide, nearly the weight of the *Titanic*.

As discussed in Chapter 12, the acetyl group in aspirin is transferred to the side chain of a serine residue that lies along the path to the active site of the cyclooxygenase component of prostaglandin H2 synthase (Section 12.5.2). In this position, the acetyl group blocks access to the active site. Thus, even though aspirin binds in the same pocket on the enzyme as salicylic acid, the acetyl group of aspirin dramatically increases its effectiveness as a drug. The account illustrates the value of screening extracts from plants and other materials that are believed to have medicinal properties for active compounds. The large number of herbal and folk medicines are a treasure trove of new drug leads.

More than 100 years ago, a fatty, yellowish material was discovered on the arterial walls of patients who had died of vascular disease. The presence of the material was termed *atheroma* from the Greek word for porridge. This material proved to be cholesterol. The Framingham heart study, initiated in 1948, documented a correlation between high blood cholesterol levels and high mortality rates from heart disease. This observation led to the notion that blocking cholesterol synthesis might lower blood cholesterol levels and, in turn, lower the risk of heart disease. Drug developers had to abandon an initial attempt at blocking the cholesterol synthesis pathway at a late step because cataracts and other side effects developed, caused by the accumulation of the insoluble substrate for the inhibited enzyme. Investigators eventually identified a more favorable target—namely, the enzyme HMG-CoA reductase (Section 26.2.1). This enzyme acts on a substrate, HMG-CoA

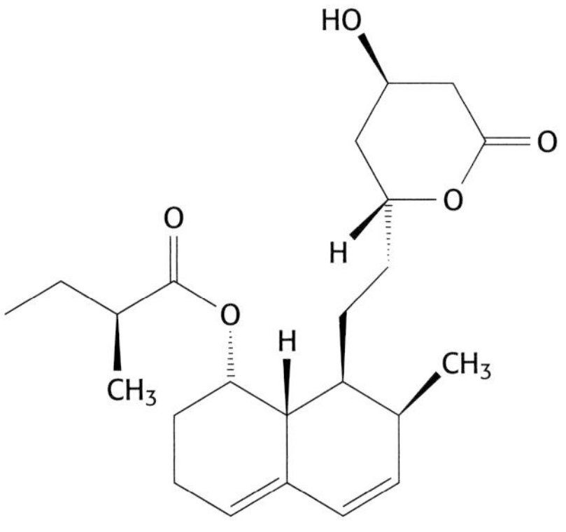

(3-hydroxy-3-methylglutaryl coenzyme A), that can be used by other pathways and is water soluble.

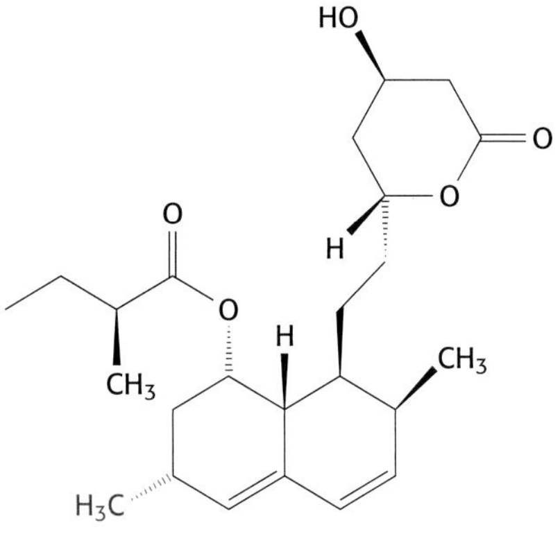

A promising natural product, compactin, was discovered in a screen of compounds from a fermentation broth from *Penicillium citrinum* in a search for antibacterial agents. In some, but not all, animal studies, compactin was found to inhibit HMG-CoA reductase and to lower serum cholesterol levels. In 1982, a new HMG-CoA reductase inhibitor was discovered in a fermentation broth from *Aspergillus cereus*. This compound, now called lovastatin, was found to be structurally very similar to compactin, bearing one additional methyl group.

In clinical trials, lovastatin significantly reduced serum cholesterol levels with few side effects. Most side effects could be prevented by treatment with mevalonate (the product of HMG-CoA reductase), indicating that the side effects were likely due to the highly effective blocking of HMG-CoA reductase. One notable side effect is muscle pain or weakness (termed *myopathy*), although its cause remains to be fully established. After many studies the Food and Drug Administration (FDA) approved lovastatin for treating high serum cholesterol levels.

A structurally related HMG-CoA reductase inhibitor was later shown to cause a statistically significant decrease in deaths due to coronary heart disease. This result validated the benefits of lowering serum cholesterol levels. Further mechanistic analysis revealed that the HMG-CoA reductase inhibitor acts not only by lowering the rate of cholesterol biosynthesis, but also by inducing the expression of the low-density-lipoprotein (LDL) receptor (Section 26.3.3). Cells with such receptors remove LDL particles from the bloodstream, and so these particles cannot contribute to atheroma.

Lovastatin and its relatives are natural products or compounds readily derived from natural products. The next step was the development of totally synthetic molecules that are more potent inhibitors of HMG-CoA reductase (Figure 18). These compounds are effective at lower dose levels, reducing side effects.

The original HMG-CoA reductase inhibitors or their precursors were found by screening libraries of natural products. More recently, drug developers have tried screening large libraries of both natural products and purely synthetic compounds prepared in the course of many drug-development programs. Under favorable circumstances, hundreds of thousands or even millions of compounds can be tested in this process, termed *high-throughput*

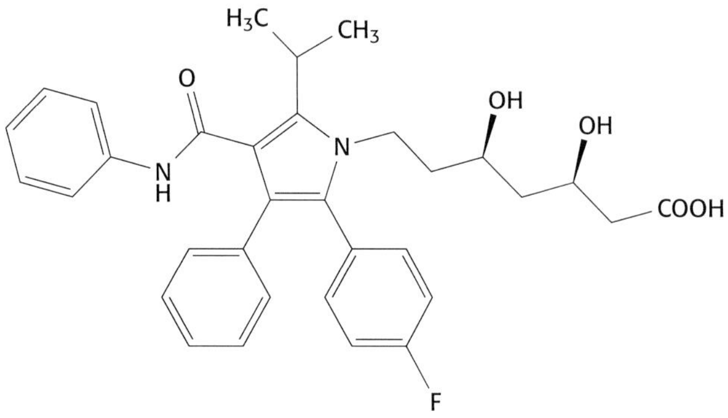

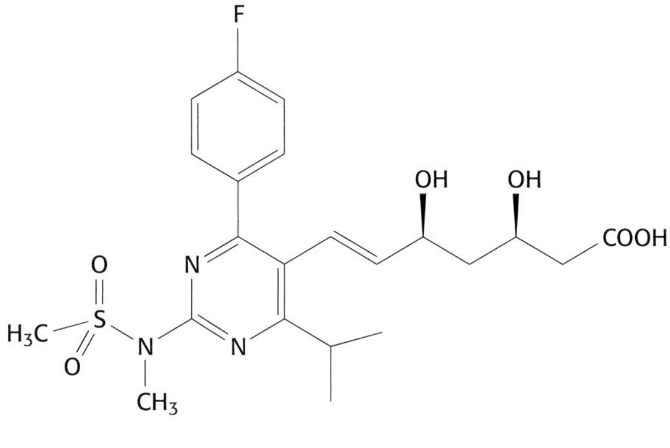

Figure 18 Synthetic statins. Atorvastatin (Lipitor®) and rosuvastatin (Crestor®) are completely synthetic drugs that inhibit HMG-CoA reductase.

screening. Compounds in these libraries can be synthesized one at a time for testing. An alternative approach is to synthesize a large number of structurally related compounds that differ from one another at only one or a few positions all at once. This approach is often termed *combinatorial chemistry*. Here, compounds are synthesized with the use of the same chemical reactions but a variable set of reactants. Suppose that a molecular scaffold is constructed with two reactive sites and that 20 reactants can be used in the first site and 40 reactants can be used in the second site. A total of $20 \times 40 = 800$ possible compounds can be produced.

A key method in combinatorial chemistry is *split-pool synthesis* (Figure 19). The method depends on solid-phase synthetic methods, first developed

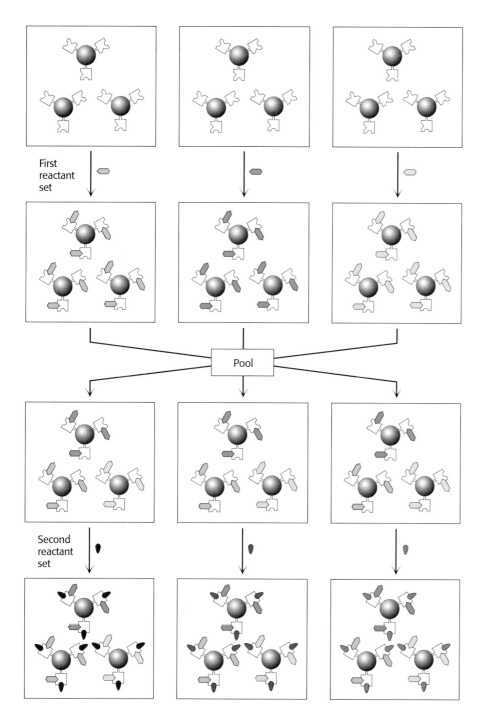

Figure 19 Split-pool synthesis. Reactions are performed on beads. Each of the reactions with the first set of reactants is performed on a separate set of beads. The beads are then pooled, mixed, and split into sets. The second set of reactants is then added. Many different compounds will be produced, but all of the compounds on a single bead will be identical.

for the synthesis of peptides (Section 4.4). Compounds are synthesized on small beads. Beads containing an appropriate starting *scaffold* are produced and divided (split) into n sets, with n corresponding to the number of building blocks to be used at one site. Reactions adding the reactants at the first site are run, and the beads are isolated by filtration. The n sets of beads are then combined (pooled), mixed, and split again into m sets, with m corresponding to the number of reactants to be used at the second site. Reactions adding these m reactants are run, and the beads are again isolated. The important result is that each bead contains only one compound, even though the entire library of beads contains many. Furthermore, although only $n + m$ reactions were run, $n \times m$ compounds are produced. With the preceding values for n and m, $20 + 40 = 60$ reactions produce $20 \times 40 = 800$ compounds. In some cases, assays can be performed directly with the compounds still attached to the bead to find compounds with desired properties (Figure 20). Alternatively, each bead can be isolated and the compound can be cleaved from the bead to produce free compounds for analysis. After an interesting compound has been identified, analytical methods of various types must be used to identify which of the $n \times m$ compounds is present.

Note that the "universe" of druglike compounds is vast. More than an estimated 10^{40} compounds are possible with molecular weights less than 750. Thus, even with "large" libraries of millions of compounds, only a tiny fraction of the chemical possibilities are present for study.

Drugs Can Be Designed on the Basis of Three-Dimensional Structural Information About Their Targets

Many drugs bind to their targets in a manner reminiscent of Emil Fischer's lock and key. Given this fact, one should be able to design a key given enough knowledge about the shape and chemical composition of the lock. In the idealized case, one would like to design a small molecule that is complementary in shape and electronic structure to a target protein so that it binds effectively to the targeted site. Despite our ability to determine three-dimensional structures rapidly, the achievement of this goal remains in the future. It is difficult to design from scratch stable compounds that have the correct shape and other properties to fit precisely

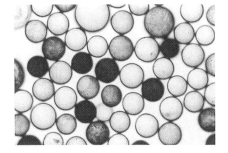

Figure 20 Screening a library of synthesized carbohydrates. A small combinatorial library of carbohydrates synthesized on the surface of 130-μm beads is screened for carbohydrates that are bound tightly by a lectin from peanuts. Beads that have such carbohydrates are darkly stained through the action of an enzyme linked to the lectin. [From R. Liang et al., *Proc. Natl. Acad. Sci. USA* 94(1997): 10554–10559; © 2004 National Academy of Sciences, USA.]

into a binding site because it is difficult to predict the structure that will best fit into a binding site. Prediction of binding affinity requires a detailed understanding of the interactions between a compound and its binding partner *and* of the interactions between the compound and the solvent when the compound is free in solution.

Nonetheless, *structure-based drug design* has proved to be a powerful tool in drug development. One of its most prominent successes has been the development of drugs that inhibit the protease from the HIV virus. Consider the development of the protease inhibitor indinavir (Crixivan®; Section 9.1.7). Two sets of promising inhibitors were discovered that had high potency but poor solubility and bioavailability. X-ray crystallographic analysis and molecular-modeling findings suggested that a hybrid molecule might have both high potency and improved bioavailability (Figure 21). The synthesized hybrid compound did show improvements but required further optimization. The structural data suggested one point where modifications could be tolerated. A series of compounds were produced and examined (Figure 22). The most active compound showed poor bioavailability, but one of the other compounds showed good bioavailability and acceptable activity. The maximum serum concentration available through oral administration was significantly higher than the levels required to suppress replication of the virus. This drug, as well as other protease inhibitors developed at about the same time, has been used in combination with other drugs to treat AIDS with much more encouraging results than had been obtained previously (Figure 23).

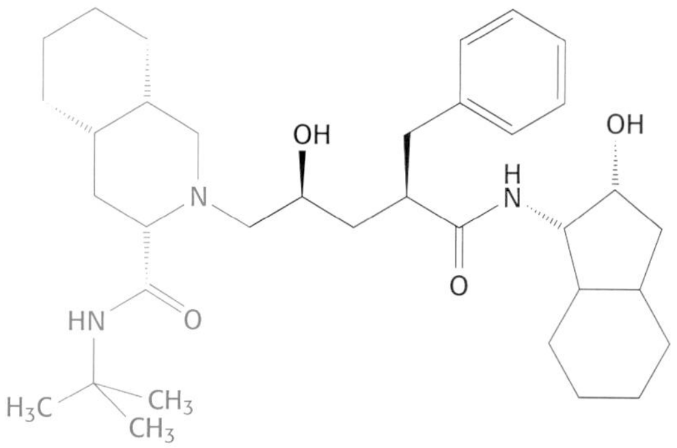

Figure 21 Initial design of an HIV protease inhibitor. This compound was designed by combining part of one compound with good inhibition activity but poor solubility (shown in red) with part of another compound with better solubility (shown in blue).

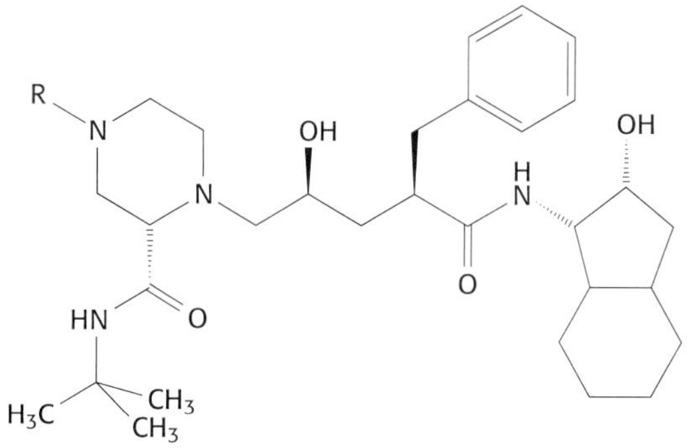

R =	IC_{50}(nmol)	log (*P*)	c_{max}(μM)
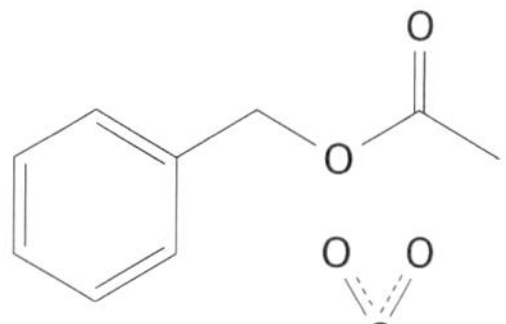 benzyl acetate	0.4	4.67	< 0.1
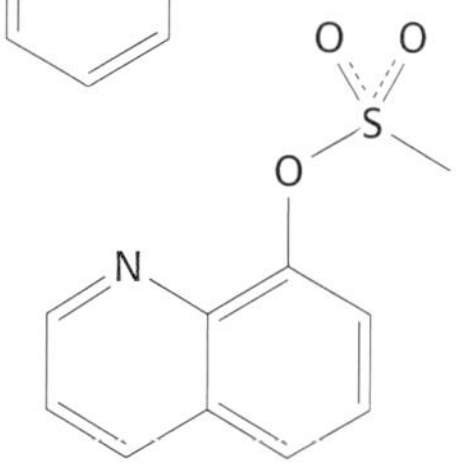 quinolinyl methanesulfonate	0.01	3.70	< 0.1
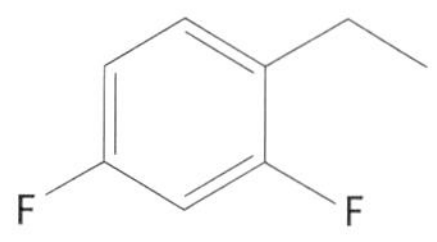 2,4-difluoro-ethylbenzene	0.3	3.69	0.7
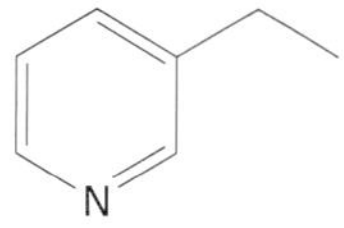 3-ethylpyridine	0.6	2.92	11

Figure 22 Compound optimization. Four compounds are evaluated for characteristics including the IC_{50} (the compound concentration required to reduce HIV replication to 50% of its maximal value), log *P*, and c_{max} (the maximal concentration of compound present) measured in the serum of dogs. The compound shown at the bottom has the weakest inhibitory power (measured by IC_{50}) but by far the best bioavailability (measured by c_{max}). This compound was selected for further development, leading to the drug indinavir (Crixivan).

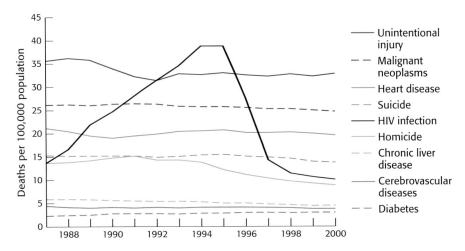

Figure 23 The effect of anti-HIV drug development. Death rates from HIV infection (AIDS) reveal the tremendous effect of HIV protease inhibitors and their use in combination with inhibitors of HIV reverse transcriptase. These are death rates from the leading causes of death among persons 24 to 44 years old in the United States. [From Centers for Disease Control.]

Aspirin targets the cyclooxygenase site in prostaglandin H2 synthase, as discussed earlier. Animal studies suggested that mammals contain not one but two distinct cyclooxygenase enzymes, both of which are targeted by aspirin. The more recently discovered enzyme, cyclooxygenase 2 (COX2), is expressed primarily as part of the inflammatory response, whereas cyclooxygenase 1 (COX1) is expressed more generally. These observations suggested that a cyclooxygenase inhibitor that was specific for COX2 might be able to reduce inflammation in conditions such as arthritis without producing the gastric and other side effects associated with aspirin.

The amino acid sequences of COX1 and COX2 were deduced from cDNA cloning studies. These sequences are more than 60% identical, clearly indicating that the enzymes have the same overall structure. Nevertheless, there are some differences in the residues around the aspirin-binding site. X-ray crystallography revealed that an extension of the binding pocket was present in COX2, but absent in COX1. This structural difference suggested a strategy for constructing COX2-specific inhibitors—namely, to synthesize compounds that had a protuberance that would fit into the pocket in the COX2 enzyme. Such compounds were designed and synthesized and then further refined to produce effective drugs familiar as Celebrex® and Vioxx® (Figure 24). The successful development of these

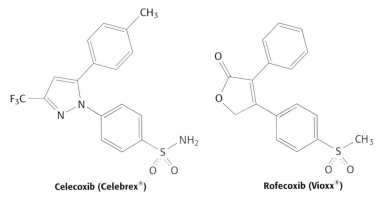

Figure 24 COX2-specific inhibitors. These compounds have protuberances (shown in red) that fit into a pocket in the COX2 isozyme but sterically clash with the COX1 isozyme.

drugs reveals the power of structure-based drug design to produce compounds specific enough to distinguish even closely related targets.

3 The Analysis of Genomes Holds Great Promise For Drug Discovery

The completion of the sequencing of the human and other genomes is a potentially powerful driving force for the development of new drugs. Genomic sequencing and analysis projects have vastly increased our knowledge of the proteins encoded by the human genome. This new source of knowledge may greatly accelerate early stages of the drug-development process or even allow drugs to be tailored to the individual patient.

Potential Targets Can Be Identified in the Human Proteome

The human genome encodes approximately 23,000 proteins, not counting the variation produced by alternative mRNA splicing and posttranslational modifications. Many of these proteins are potential drug targets, in particular those that are enzymes or receptors and have significant biological effects when activated or inhibited. Several large protein families are particularly rich sources of targets. For example, the human genome includes genes for more than 500 protein kinases that can be recognized by comparing the deduced amino acid sequences. One of them, Bcr-Abl kinase, is known to contribute to leukemias and is the target of the drug imatinib mesylate (Gleevec®; Section 15.5.1). Some of the other protein kinases undoubtedly play central roles in particular cancers as well. Similarly, the human genome encodes approximately 800 7TM receptors (Section 15.1) of which approximately 350 members are odorant receptors. Many of the remaining family members are potential drug targets. Some of them are already targets for drugs, such as the beta-blocker atenolol, which targets the β-adrenergic receptor, and the antiulcer medication ranitidine (Zantac®). The latter compound is an antagonist of the histamine H2 receptor, a 7TM receptor that participates in the control of gastric acid secretion.

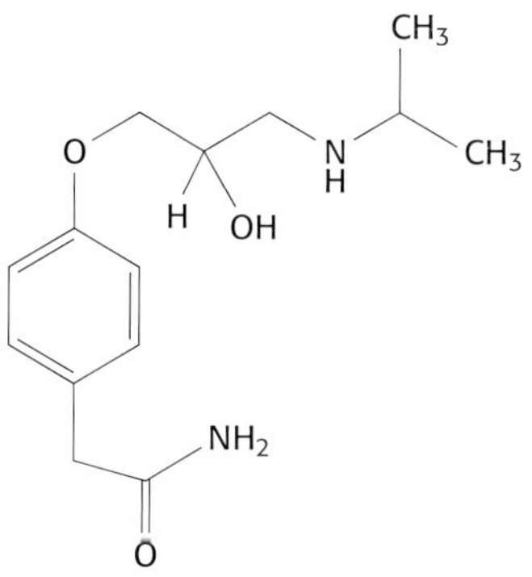

Atenolol

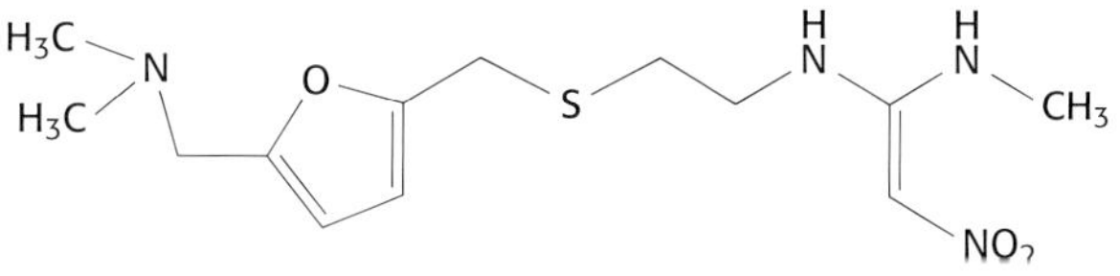

Ranitidine

Novel proteins that are not part of large families already supplying drug targets can be more readily identified through the use of genomic information.

There are a number of ways to identify proteins that could serve as targets of drug-development programs. One way is to look for changes in expression patterns, protein localization, or posttranslational modifications in cells from disease-afflicted organisms. Another is to perform studies of tissues or cell types in which particular genes are expressed. Analysis of the human genome should increase the number of actively pursued drug targets by a factor of an estimated two or more.

Animal Models Can Be Developed to Test the Validity of Potential Drug Targets

The genomes of a number of model organisms have now been sequenced. The most important of these genomes for drug development is that of the mouse. Remarkably, the mouse and human genomes are approximately 70% identical in sequence, and more than 98% of all human genes have recognizable mouse counterparts. Mouse studies provide drug developers with a powerful tool—the ability to disrupt ("knock out") specific genes in the mouse (Section 6.3.5). If disruption of a gene has a desirable effect, then the product of this gene is a promising drug target. The utility of this approach has been demonstrated retrospectively. For example, disruption of the gene for the α subunit of the H^+-K^+ ATPase, the key protein for secreting acid into the stomach, produces mice with less acid in their stomachs. The stomach pH of such mice is 6.9 in circumstances that produce a stomach pH of 3.2 in their wild-type counterparts. This protein is the target of the drugs omeprazole (Prilosec®) and lansoprazole (Prevacid® and Takepron®), used for treating gastric-esophageal reflux disease.

Omeprazole

Lansoprazole

Several large-scale efforts are underway to generate hundreds or thousands of mouse strains, each having a different gene disrupted. The phenotypes of these mice are a good indication of whether the protein encoded by a disrupted gene is a promising drug target. This approach allows drug developers to evaluate potential targets without any preconceived notions regarding physiological function.

Potential Targets Can Be Identified in the Genomes of Pathogens

Human proteins are not the only important drug targets. Drugs such as penicillin and HIV protease inhibitors act by targeting proteins within a pathogen. The genomes of hundreds of pathogens have now been sequenced, and these genome sequences can be mined for potential targets.

New antibiotics are needed to combat bacteria that are resistant to many existing antibiotics. One approach seeks proteins essential for cell survival that are conserved in a wide range of bacteria. Drugs that inactivate such proteins are expected to be broad-spectrum antibiotics, useful for treating infections from any of a range of different bacteria. One such protein is peptide deformylase, the enzyme that removes formyl groups that

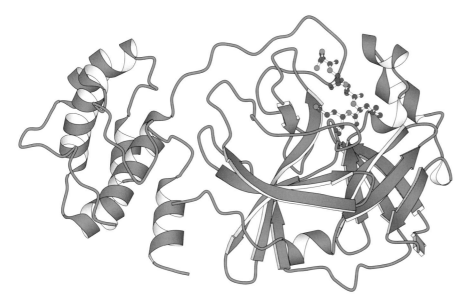

Figure 25 Emerging drug target. The structure of a protease from the coronavirus that causes SARS (severe acute respiratory syndrome) is shown bound to an inhibitor. This structure was determined less than a year after the identification of the virus.

are present at the amino termini of bacterial proteins immediately after translation (Section 29.3.5).

Alternatively, a drug may be needed against a specific pathogen. A recent example of such a pathogen is the organism responsible for severe acute respiratory syndrome (SARS). Within one month of the recognition of this emerging disease, investigators had isolated the virus that causes the syndrome, and, within weeks, its 29,751 base genome had been completely sequenced. This sequence revealed the presence of a gene encoding a viral protease, known to be essential for viral replication from studies of other members of the coronavirus family to which the SARS virus belongs. Drug developers are already at work seeking specific inhibitors of this protease (Figure 25).

Genetic Differences Influence Individual Responses to Drugs

Many drugs are not effective in everyone, often because of genetic differences between people. Nonresponding persons may have slight differences in either a drug's target molecule or proteins taking part in drug transport and metabolism. The goal of the emerging fields of pharmacogenetics and pharmacogenomics is to design drugs that either act more consistently from person to person or are tailored to individual persons with particular genotypes.

Drugs such as metoprolol that target the β1-adrenergic receptor are popular treatments for hypertension.

Metoprolol

But some people do not respond well. Two variants of the gene coding for the β1-adrenergic receptor are common in the American population.

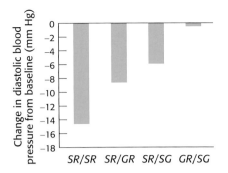

Figure 26 Phenotype–genotype correlation. Average changes in diastolic blood pressure on treatment with metoprolol. Persons with two copies of the most common ($S_{49}R_{389}$) allele showed significant decreases in blood pressure. Those with one variant allele (*GR* or *SG*) showed more modest decreases, and those with two variant alleles (*GR/SG*) showed no decrease. [From J. A. Johnson et al., *Clin. Pharmacol. Ther.* 74(2003): 44–52.]

The most common allele has serine in position 49 and arginine in position 389. In some persons, however, glycine replaces one or the other of these residues. In studies, participants with two copies of the most common allele responded well to metoprolol: their daytime diastolic blood pressure was reduced by 14.7 ± 2.9 mm Hg on average. In contrast, participants with one variant allele showed a smaller reduction in blood pressure, and the drug had no significant effect on participants with two variant alleles (Figure 26). These observations suggest the potential utility of genotyping individual persons at these positions. One could then predict whether or not treatment with metoprolol or other β-blockers is likely to be effective.

Given the importance of ADME and toxicity properties in determining drug efficacy, it is not surprising that variations in proteins participating in drug transport and metabolism can alter a drug's effectiveness. An important example is the use of thiopurine drugs such as 6-thioguanine, 6-mercaptopurine, and azothioprine to treat diseases including leukemia, immune disorders, and inflammatory bowel disease.

6-Thioguanine   **6-Mercaptopurine**   **Azathioprine**

A minority of patients who are treated with these drugs show signs of toxicity at doses that are well tolerated by most patients. These differences between patients are due to rare variations in the gene encoding the xenobiotic-metabolizing enzyme thiopurine methyltransferase, which adds a methyl group to sulfur atoms.

6-Mercaptopurine + *S*-adenosylmethionine ⇌ [Thiopurine methyltransferase] + *S*-adenosylhomocysteine + H⁺

The variant enzyme is less stable. Patients with these variant enzymes can build up toxic levels of the drugs if appropriate care is not taken. Thus, genetic variability in an enzyme participating in drug metabolism plays a large role in determining the variation in the tolerance of different persons to particular drug levels. Many other drug-metabolism enzymes and drug-transport proteins have been implicated in controlling individual reactions to specific drugs. The identification of the genetic factors will allow a deeper understanding of why some drugs work well in some persons but poorly in others. In the future, doctors may examine a patient's genotype with respect to these genes to help plan drug-therapy programs.

4 The Development of Drugs Proceeds Through Several Stages

The FDA requires that drug candidates be demonstrated to be effective and safe before they may be used in human beings on a large scale. This requirement is particularly true for drug candidates that are to be taken by people who are relatively healthy. More side effects are acceptable for drug candidates intended to treat significantly ill patients such as those with serious forms of cancer, where there are clear, unfavorable consequences for not having an effective treatment.

Clinical Trials Are Time Consuming and Expensive

Clinical trials test the effectiveness and potential side effects of a candidate drug before it is approved by the FDA for general use. These trials proceed in at least three phases (Figure 27). In phase 1, a small number (usually from 10 to 100) of healthy volunteers take the drug for an initial study of safety. These volunteers are given a range of doses and are monitored for signs of toxicity. The efficacy of the drug candidate is not specifically evaluated.

In phase 2, the efficacy of the drug candidate is tested in a small number of persons who might benefit from the use of the drug. Further data regarding the safety of the drug candidate are obtained. Such trials are often controlled and double-blinded. In a controlled study, subjects are divided randomly into two groups. Subjects in the treatment group are given the treatment under investigation. Subjects in the control group are given either a placebo—that is, a treatment such as sugar pills known to not have intrinsic value—or the best standard treatment available, if withholding treatment altogether would be unethical. In a double-blinded study, neither the subjects nor the researchers know which subjects are in the treatment group and which are in the control group. A double-blinded study prevents bias in the course of the trial. When the trial has been completed, the assignments of the subjects into treatment and control groups are unsealed and the results for the two groups are compared. A variety of doses are often investigated in phase 2 trials to determine which doses appear to be free of serious side effects and which doses appear to be effective.

One should not underestimate the power of the placebo effect—that is, the tendency to perceive improvement in a subject who believes that he or she is receiving a potentially beneficial treatment. In a study of arthroscopic surgical treatment for knee pain, subjects who were led to believe that they had received surgery through the use of videotapes and other means showed the same level of improvement, on average, as subjects who were actually operated on.

In phase 3, similar studies are performed on a larger population. This phase is intended to more firmly establish the efficacy of the drug candidate and to detect side effects that may develop in a small percentage of the subjects who receive treatment. Thousands of subjects may participate in a typical phase 3 study.

Clinical trials can be extremely costly. Hundreds or thousands of patients must be recruited and monitored for the duration of the trial.

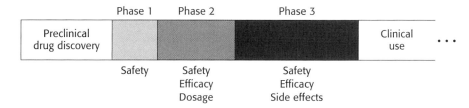

Figure 27 Clinical-trial phases. Clinical trials proceed in phases examining safety and efficacy in increasingly large groups.

Many physicians, nurses, clinical pharmacologists, statisticians, and others participate in the design and execution of the trial. Costs can run from tens of millions to hundreds of millions of dollars. Extensive records must be kept, including documentation of any adverse reactions. These data are compiled and submitted to the FDA. The full cost of developing a drug is currently estimated to be from $400 million to $800 million.

Even once a drug has been approved and is in use, difficulties can arise. For example, rofecoxib (Vioxx®) was withdrawn from the market after significant cardiac side effects were detected during additional clinical trials.

The Evolution of Drug Resistance Can Limit the Utility of Drugs for Infectious Agents and Cancer

Many drugs are used for long periods of time without any loss of effectiveness. However, in some cases, particularly for the treatment of infectious diseases or of cancer, drug treatments that were initially effective become less effective. In other words, the disease becomes resistant to the drug therapy. Why does this occur? Infectious diseases and cancer have a common feature— namely, that an affected person contains many cells (or viruses) that can mutate and reproduce. These conditions are necessary for evolution to take place. Thus, an individual microorganism or cancer cell may by chance have a genetic variation that makes it more suitable for growth and reproduction in the presence of the drug than is the population of microorganisms or cancer cells at large. These microorganisms or cells are more fit than others in their population, and they will tend to take over the population. As the selective pressure due to the drug is continually applied, the population of microorganisms or cancer cells will tend to become more and more resistant to the presence of the drug. Note that resistance can develop by a number of mechanisms.

The HIV protease inhibitors discussed earlier provide an important example of the evolution of drug resistance. Retroviruses are very well suited to this sort of evolution because reverse transcriptase carries out replication without a proofreading mechanism. In a genome of approximately 9750 bases, each possible single point mutation is estimated to appear in a virus particle more than 1000 times per day in each infected person. Many multiple mutations also occur. Most of these mutations either have no effect or are detrimental to the virus. However, a few of the mutant virus particles encode proteases that are less susceptible to inhibition by the drug. In the presence of an HIV protease inhibitor, these viruses will tend to replicate more effectively than the population at large. Over time, these viruses will come to dominate the population and the virus population will become resistant to the drug.

Pathogens may become resistant to antibiotics by completely different mechanisms. Some pathogens contain enzymes that inactivate or degrade specific antibiotics. For example, many organisms are resistant to β-lactams such as penicillin because they contain β-lactamase enzymes. These enzymes hydrolyze the β-lactam ring and render the drugs inactive.

Many of these enzymes are encoded in plasmids, small circular pieces of DNA often carried by bacteria. Many plasmids are readily transferred from one bacterial cell to another, transmitting the capability for antibiotic resistance. Plasmid transfer thus contributes to the spread of antibiotic resistance, a major health-care challenge. On the other hand, plasmids have been harnessed for use in recombinant DNA methods (Section 6.2.1).

Drug resistance commonly emerges in the course of cancer treatment. Cancer cells are characterized by their ability to grow rapidly without the constraints that apply to normal cells. Many drugs used for cancer chemotherapy inhibit processes that are necessary for this rapid cell growth. However, individual cancer cells may accumulate genetic changes that mitigate the effects of such drugs. These altered cancer cells will tend to grow more rapidly than others and will become dominant within the cancer-cell population. This ability of cancer cells to mutate quickly has posed a challenge to one of the major breakthroughs in cancer treatment: the development of inhibitors for proteins specific to cancer cells present in certain leukemias (Section 15.5.1). For example, tumors became undetectable in patients treated with imatinib mesylate, which is directed against the Bcr-Abl protein kinase. Unfortunately, the tumors of many of the patients treated with imatinib mesylate recur after a period of years. In many of these cases, mutations have altered the Bcr-Abl protein so that it is no longer inhibited by the concentrations of imatinib mesylate used in therapy.

Cancer patients often take multiple drugs concurrently in the course of chemotherapy, and in many cases cancer cells become simultaneously resistant to many or all of them. This multiple-drug resistance can be due to the proliferation of cancer cells that overexpress a number of ABC transporter proteins that pump drugs out of the cell (Section 13.3). Thus, cancer cells can evolve drug resistance by overexpressing normal human proteins or by modifying proteins responsible for the cancer phenotype.

Summary

1 The Development of Drugs Presents Huge Challenges

Most drugs act by binding to enzymes or receptors and modulating their activities. To be effective, drugs must bind to these targets with high affinity and specificity. However, even most compounds with the desired affinity and specificity do not make suitable drugs. Most compounds are poorly absorbed or rapidly excreted from the body or they are modified by metabolic pathways that target foreign compounds. Consequently, when taken orally, these compounds do not reach their targets at appropriate concentrations for a sufficient period of time. A drug's properties related to its absorption, distribution, metabolism, and excretion are called ADME properties. Oral bioavailability is a measure of a drug's ability to be absorbed; it is the ratio of the peak concentration of a compound given orally to the peak concentration of the same dose directly injected. The structure of a compound can affect its bioavailability in complicated ways, but generalizations called Lipinski's rules provide useful guidelines. Drug metabolism pathways include oxidation by cytochrome P450 enzymes (phase I metabolism) and conjugation to glutathione, glucuronic acid, and sulfate (phase II metabolism). A compound may also not be a useful drug because it is toxic, either because it modulates the target molecule too effectively or because it also binds to proteins other than the target. The liver and kidneys play central roles in drug metabolism and excretion.

2 Drug Candidates Can Be Discovered by Serendipity, Screening, or Design

Many drugs have been discovered by serendipity—that is, by chance observation. The antibiotic penicillin is produced by a mold that accidentally contaminated a culture dish, killing nearby bacteria. Drugs such as chlorpromazine and sildenafil were discovered to have beneficial effects on human physiology that were completely different from those expected. The cholesterol-lowering statin drugs were developed after large collections of compounds were screened for potentially interesting activities. Combinatorial chemistry methods have been developed to generate large collections of chemically related yet diverse compounds for screening. In some cases, the three-dimensional structure of a drug target is available and can be used to aid the design of potent and specific inhibitors. Examples of drugs designed in this manner are the HIV protease inhibitors indinavir and cyclooxygenase 2 inhibitors such as celecoxib.

3 The Analysis of Genomes Holds Great Promise for Drug Discovery

The human genome encodes approximately 23,000 proteins, and many more if derivatives due to alternative mRNA splicing and posttranslational modification are included. The genome sequences can be examined for potential drug targets. Large families of proteins known to participate in key physiological processes such as the protein kinases and 7TM receptors each yielded several targets for which drugs have been developed. The genomes of model organisms also are useful for drug-development studies. Strains of mice with particular genes disrupted have been useful in validating certain drug targets. The genomes of bacteria, viruses, and parasites encode many potential drug targets that can be exploited owing to their important functions and their differences from human proteins, minimizing the potential for side effects. Genetic differences between individual people can be examined and correlated with differences in responses to drugs, potentially aiding both clinical treatments and drug development.

4 The Development of Drugs Proceeds Through Several Stages

Before compounds can be given to human beings as drugs, they must be extensively tested for safety and efficacy. Clinical trials are performed in stages, first testing safety, then safety and efficacy in a small population, and finally safety and efficacy in a larger population to detect rarer adverse effects. Largely due to the expenses associated with clinical trials, the cost of developing a new drug has been estimated to be as much as $800 million. Even when a drug has been approved for use, complications can arise. With infectious diseases and cancer, patients often develop resistance to a drug after it has been used for some period of time, because variants of the disease agent that are less susceptible to the drug arise and replicate, even when the drug is present.

Key Terms

ADME (p. 4)
apparent dissociation constant, K_d^{app} (p. 3)
atheroma (p. 13)
blood–brain barrier (p. 6)
combinatorial chemistry (p. 15)
compartment (p. 6)
conjugation (p. 6)
dissociation constant, K_d (p. 3)
drug metabolism (p. 6)
enterohepatic cycling (p. 9)
first-pass metabolism (p. 8)

glomerulus (p. 8)
high-throughput screening (p. 14)
inhibition constant, K_i (p. 3)
ligand (p. 2)
Lipinski's rules (p. 4)
myopathy (p. 14)
oral bioavailability (p. 4)
oxidation (p. 6)
phase I transformation (p. 8)
phase II transformation (p. 8)
side effects (p. 3)
split-pool synthesis (p. 15)
structure-based drug design (p. 17)
therapeutic index (p. 9)
xenobiotic compounds (p. 6)

Selected Readings

Books
Hardman, J. G., Limbird, L. E., and Gilman, A. G. 2001. *Goodman and Gilman's The Pharmacological Basis of Therapeutics* (10th ed.). McGraw-Hill Professional.

Levine, R. R., and Walsh, C. T. 2004. *Levine's Pharmacology: Drug Actions and Reactions* (7th ed.). Taylor and Francis Group.

Silverman, R. B. 2004. *Organic Chemistry of Drug Design and Drug Action*. Academic Press.

ADME and Toxicity
Caldwell, J., Gardner, I., and Swales, N. 1995. An introduction to drug disposition: The basic principles of absorption, distribution, metabolism, and excretion. *Toxicol. Pathol.* 23:102–114.

Lee, W., and Kim, R. B. 2004. Transporters and renal drug elimination. *Annu. Rev. Pharmacol. Toxicol.* 44:137–166.

Lin, J., Sahakian, D. C., de Morais, S. M., Xu, J. J., Polzer, R. J., and Winter, S. M. 2003. The role of absorption, distribution, metabolism, excretion and toxicity in drug discovery. *Curr. Top. Med. Chem.* 3:1125–1154.

Poggesi, I. 2004. Predicting human pharmacokinetics from preclinical data. *Curr. Opin. Drug Discov. Devel.* 7:100–111.

Case Histories
Flower, R. J. 2003. The development of COX2 inhibitors. *Nat. Rev. Drug Discov.* 2:179–191.

Tobert, J. A. 2003. Lovastatin and beyond: The history of the HMG-CoA reductase inhibitors. *Nat. Rev. Drug Discov.* 2:517–526.

Vacca, J. P., Dorsey, B. D., Schleif, W. A., Levin, R. B., McDaniel, S. L., Darke, P. L., Zugay, J., Quintero, J. C., Blahy, O. M., Roth, E., et al. 1994. L-735,524: An orally bioavailable human immunodeficiency virus type 1 protease inhibitor. *Proc. Natl. Acad. Sci. USA* 91:4096–4100.

Wong, S., and Witte, O. N. 2004. The BCR-ABL story: Bench to bedside and back. *Annu. Rev. Immunol.* 22:247–306.

Structure-Based Drug Design
Kuntz, I. D. 1992. Structure-based strategies for drug design and discovery. *Science* 257:1078–1082.

Dorsey, B. D., Levin, R. B., McDaniel, S. L., Vacca, J. P., Guare, J. P., Darke, P. L., Zugay, J. A., Emini, E. A., Schleif, W. A., Quintero, J. C., et al. 1994. L-735,524: The design of a potent and orally bioavailable HIV protease inhibitor. *J. Med. Chem.* 37:3443–3451.

Chen, Z., Li, Y., Chen, E., Hall, D. L., Darke, P. L., Culberson, C., Shafer, J. A., and Kuo, L. C. 1994. Crystal structure at 1.9-Å resolution of human immunodeficiency virus (HIV) II protease complexed with L-735,524, an orally bioavailable inhibitor of the HIV proteases. *J. Biol. Chem.* 269:26344–26348.

Combinatorial Chemistry
Baldwin, J. J. 1996. Design, synthesis and use of binary encoded synthetic chemical libraries. *Mol. Divers.* 2:81–88.

Burke, M. D., Berger, E. M., and Schreiber, S. L. 2003. Generating diverse skeletons of small molecules combinatorially. *Science* 302:613–618.

Edwards, P. J., and Morrell, A. I. 2002. Solid-phase compound library synthesis in drug design and development. *Curr. Opin. Drug Discov. Devel.* 5:594–605.

Genomics
Zambrowicz, B. P., and Sands, A. T. 2003. Knockouts model the 100 best-selling drugs: Will they model the next 100? *Nat. Rev. Drug Discov.* 2:38–51.

Salemme, F. R. 2003. Chemical genomics as an emerging paradigm for postgenomic drug discovery. *Pharmacogenomics.* 4:257–267.

Michelson, S., and Joho, K. 2000. Drug discovery, drug development and the emerging world of pharmacogenomics: Prospecting for information in a data-rich landscape. *Curr. Opin. Mol. Ther.* 2:651–654.

Weinshilboum, R., and Wang, L. 2004. Pharmacogenomics: Bench to bedside. *Nat. Rev. Drug Discov.* 3:739–748.

Problems

1. *Routes to discovery.* For each of the following drugs, indicate whether the physiological effects of the drug were known before or after the target was identified.

   (a) Penicillin
   (b) Sildenafil (Viagra®)
   (c) Rofecoxib (Vioxx®)
   (d) Atorvastatin (Lipitor®)
   (e) Aspirin
   (f) Indinavir (Crixivan®)

2. *Lipinski's rules.* Which of the following compounds satisfy all of Lipinski's rules? [Log(P) values are given in parentheses.]

   (a) Atenolol (0.23)
   (b) Sildenafil (3.18)
   (c) Indinavir (2.78)

3. *Calculating log tables.* Considerable effort has been expended to develop computer programs that can estimate log(P) values entirely on the basis of chemical structure. Why would such programs be useful?

4. *An ounce of prevention.* Legislation has been proposed that would require that acetaminophen tablets also include N-acetylcysteine. Speculate about the role that this additive would serve.

5. *Drug interactions.* As noted in the chapter, coumadin can be a very dangerous drug because too much can cause uncontrolled bleeding. Persons taking coumadin must be careful about taking other drugs, particularly those that bind to albumin. Propose a mechanism for this drug–drug interaction.

6. *Find the target.* Trypanosomes are unicellular parasites that cause sleeping sickness. During one stage of their life cycle, these organisms live in the bloodstream and derive all of their energy from glycolysis, which takes place in a specialized organelle called a glycosome inside the parasite. Propose potential targets for treating sleeping sickness. What are some potential difficulties with your approach?

Mechanism Problem

7. *Variations on a theme.* The metabolism of amphetamine by cytochrome P450 enzymes results in the conversion shown here. Propose a mechanism and indicate any additional products.

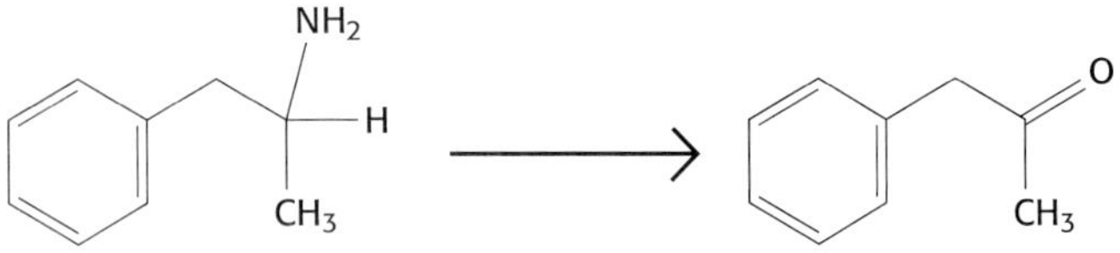

Data Interpretation Problem

8. *HIV protease inhibitor design.* Compound A is one of a series that were designed to be potent inhibitors of HIV protease.

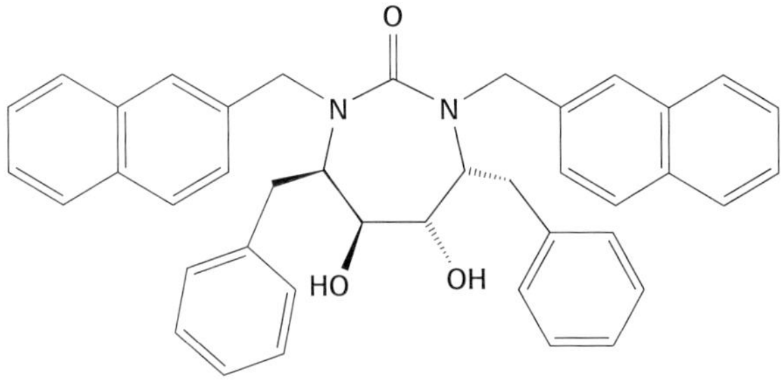

Compound A was tested by using two assays: (1) direct inhibition of HIV protease in vitro and (2) inhibition of viral RNA production in HIV infected cells, a measure of viral replication.

The results of these assays are shown here. The HIV protease activity is measured with a substrate peptide present at a concentration equal to its K_M value.

Compound A (nM)	HIV protease activity (arbitrary units)
0	11.2
0.2	9.9
0.4	7.4
0.6	5.6
0.8	4.8
1	4.0
2	2.2
10	0.9
100	0.2

Compound A (μM)	Viral RNA production (arbitrary units)
0	760
1.0	740
2.0	380
3.0	280
4.0	180
5.0	100
10	30
50	20

Estimate the values for the K_I of compound A in the protease activity assay and for its IC_{50} in the viral RNA production assay.

Treating rats with the relatively high oral dose of 20 mg kg^{-1} results in a maximum concentration of compound A of 0.4 μM. On the basis of this value, do you expect compound A to be effective in preventing HIV replication when taken orally?